WHY
THE UNIVERSE EXISTS

A SHORT, NON-TECHNICAL ESSAY
FOR THE GENERAL READER

JAMES D ROBINSON

EOCRANTIS

First Published in 2021
by
Eocrantis Publishing
10 Bewick Crescent
Newton Aycliffe
Co Durham
DL5 5LQ
England

Cover Design : James D Robinson
Cover image : WikiImages via Pixabay

ISBN : 978-1-9989928-1-2

Note : Source of illustration on inner title page : wikipedia
https:// en.wikipedia.org/wiki/Universe#/media/File: Flammarion

CONTENTS

ABOUT THE AUTHOR

James D Robinson was born in Darlington, Co Durham, England, in 1949. He studied traditional b/w photo printing and photographic portraiture at the Darlington College of Technology in the 1980's though he has many other interests which, in addition to writing, include: drawing, graphic design, and in the academic arena, cosmology, for which he retains an amateur status.

His photo work is represented in the permanent photo collections of the UK's: National Portrait Gallery, and The Picture Library of the Royal Society, and he is the creator of a time capsule; now stored in the Archive Department of the UK's: Durham County Record Office, where it will remain unopened until the 21st of March, 2317, making it a time capsule - at the present time of writing - with the 5th longest period of closure on record.

Note : The text in this publication is presented
without indents - by design and not error.

OTHER BOOKS by JAMES D ROBINSON

THE AUTHENTICATED METEORIC FALLS of THE BRITISH ISLES : 2009
THE PSEUDO METEORIC EVENTS OF THE BRITISH ISLES : 2010
THIS & THAT : A PHOTOGRAPHIC ALBUM : 2012
THE DAY THE BRIDGES FELL : 2013
HISTORICA PERSONICA : PHOTOGRAPHIC PORTRAITS CAPTURED AT LIVING HISTORY EVENTS : 2013
PHOTOWORKS : A MULTIVERSE OF IMAGES : 2014
THE POSTER DESIGNS OF JAMES D ROBINSON : 2017
MUSINGS ON THE COSMOS : 2018
KING THINKALOT'S ARCHIVE : TALES FROM THE TIME OF MAGIC 2019
THE NON ZERO UNIVERSE : A NEW LOOK AT THE LOGIC OF A COSMOLOGICAL EVERMORE : 2020
A TALE FROM THE FOREST : WELCOME TO THE NEIGHBOURHOOD 2020
TRANSITS : A TIME TRAVEL ODYSSEY : 2021

His work has also been published in :

THE JOURNAL OF THE ROYAL PHOTOGRAPHIC SOCIETY
POPULAR ASTRONOMY MAGAZINE
AMATEUR PHOTOGRAPHER MAGAZINE
WHAT DIGITAL CAMERA MAGAZINE
NORTH EAST LIFE MAGAZINE
SKERMISH MAGAZINE
& THE BEST OF BRITISH MAGAZINE

AUTHOR'S NOTE

Although this essay relates to the science of Cosmology, no in-depth knowledge of that subject is required by the reader in order to understand the text as presented herein.

"Nothing comes from Nothing"

Parmenides of Elea : Greek Philosopher : 6th Cent, BC.

THE GREAT MYSTERY

There is a general consensus throughout modern academia, that the Universe in which we live had a beginning. You will find it stated as such in many books, magazine and newspaper articles, as well as in tv and radio broadcasts and internet sources.
The assertion is that around 14 billion years ago, a cosmological event - now commonly referred to as 'The Big Bang event - suddenly took place, for no accountable reason, which brought the Universe into existence, and with it the emergence of time, space and reality, since before the birth of the Universe there was not even empty space.
This view came to the fore in the late 1920's following the presentation of astronomical evidence by the American astronomer Edwin Hubble, which seemed to suggest that the universe was *expanding*,[1] a revelation which rocked the scientific community at the time, since before the 'model' of the Big Bang came to prominence, the opinion amongst many

people was that the Universe was neither expanding or contracting, and was thus considered as a *static* construct, although the stars themselves could move about randomly.

It was also believed that our own galaxy, The Milky Way, was the one and only galaxy, that is to say, there was no general awareness that other galaxies existed beyond our own, although they had been observed but were mistakenly thought to be clouds of interstellar gas, otherwise known as *Nebula,* which is the Latin word for *Clouds.*

In any event, once the notion of an expanding Universe took hold, it was realised that if the rate of expansion could be calculated, then the age of the Universe would be known by imagining the expansion in reverse, and that basic assumption has been held ever since.

Another realisation of reversing the expansion was that at the moment the Big Bang event occurred, everything in the Universe must have been packed into a minute volume of space, with a temperature and pressure of astronomical proportions. That assumes of course, that no other event of any kind happened along the way.

If so, then it seems fair to ask : where did all this *substance* of the newly born Universe come from, what is it expanding into, and why did it suddenly start to expand in the first place?

With respect to the Big Bang model, these questions

still remain to be answered satisfactorily, because as things stand, the most common response you will receive from the scientific community (with respect to the first question) is that the Universe simply sprang into existence, in an instant, from *absolute nothingness,* by a process which is referred to as *spontaneous creation,* whilst those who side with the religious community will simply put everything down to the handiwork of God.

Whether you are prepared to accept either of the above explanations or an alternative, will, however, essentially depend on how you respond to two fundamental questions, viz: Do you believe in God? and: Do you favour a *finite* or *infinite* Universe?

For the present author, who is non-religious, and defines the Universe as *That which has, does and always will contain everything in existence,* the dismissal of the religious viewpoint is therefore, self explanatory, whilst the dismissal of the current Big Bang scientific viewpoint is based on the following:

When we speak of *absolute nothingness,* it is important to appreciate that what it means, in its truest sense, is the absence of everything you can imagine. So, there can be no matter, no energy, no time, no space, no dimensions of any kind and ultimately, therefore, no reality.

Logically, this clearly creates a problem with respect

to the Big Bang model since if there is *no time,* then there can never be a *when,* or to put it another way, there can never be a moment of creation. And if there is *no space,* then there can never be a *somewhere* for any kind of an event to occur.

We should also bear in mind that even a so-called *spontaneous creation* requires a certain amount of time to occur, since you cannot have an occurrence of zero time, and it would also need a source of energy to undergo the transformation.

The plain fact is, the only rational conclusions that can be arrived at in view of the above related factors are 1) that a spontaneous creation event from a state of absolute nothingness is an impossibility, and 2) that there is *no such thing* as absolute *nothingness.* It can only ever be conceived as an abstraction, since it is devoid of all physical dimensions.

The above need not eliminate the notion of some kind of Big Bang event of course, as such could still have taken place within a wider Universe that pre-existed the Big Bang event, indeed, if we are to accept all the astronomical evidence which strongly suggest there *was* some kind of Big Bang event around 14 billion years ago, then the latter seems a far more likely scenario than the notion that everything sprang from absolute nothingness.

The other outcome of the notion that *there is no such thing as absolute nothingness,* is that it leads

This circle represents the Universe of today.
It thus has mass and energy, plus dimensions in both time and space.

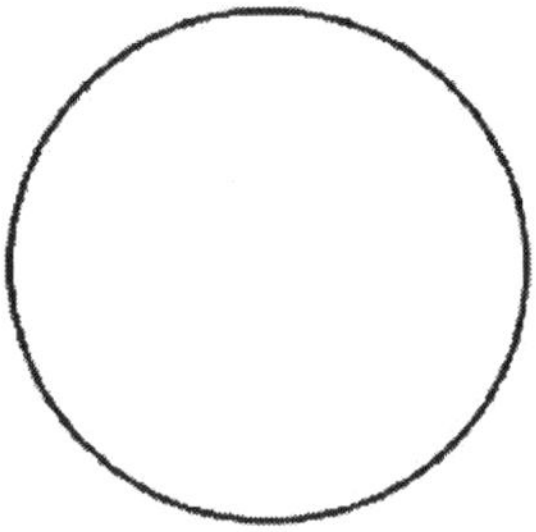

This circle represents a Universe devoid of all that exists
except its dimensions in time and space.

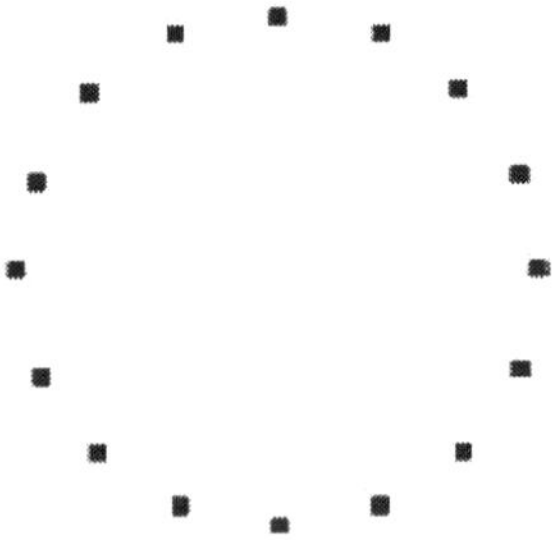

This circle is purely a concept to show that absolute nothingness
is Less than empty space, hence the circle is now reduced to dots whilst
the blank space signifies that all things are absent

us to conclude that there must *always* be a *somethingness,* as evidenced by the fact that the Universe exists, and must, therefore, have always existed and will always exist, in one form or another. In other words, the Universe is eternal.

But we can add more to it than that, for if there is no such thing as absolute nothingness, then the Universe is also *infinite* in dimension, which would solve the issue of what our *observable* Universe (the part we can see) is expanding into, viz : the space of a wider Universe where the expansion of our observable Universe is simply a localised event, and thus possibly only one of many, within a much wider Universe that is, *overall,* not expanding, since if it is infinite in size, it cannot get any bigger, it simply exists without end.

Such a concept may be difficult to fully comprehend because we are naturally inclined to visualise objects with discernible boundaries, but for a Universe where there is no absolute nothingness, there can be no ultimate boundary, and therefore no specific centre of the Universe, because from whatever point you choose, they all reach out to infinity.

So, how did all the stuff within the Universe come to be there in the first place? This is the topic of the next chapter.

THE REALITY OF REALITY

There is a simple way to resolve the question: where did all the stuff in the Universe come from? and that is to imagine you are building the Universe from scratch.

So, how might this be achieved.

Well, obviously, the first thing you would need to do is to give the Universe a dimension, and a dimension of *infinite* size at that, since it has already been shown logically: *there is no such thing as absolute nothingness,* hence space never stops.

That allows us to have a volume of space of infinite size, but one which is empty of all physical things. It does, however, have one additional attribute, and that is *time,* because time and space are inseparable, and nothing can exist for zero time.

But if this was all you ever had, then nothing would ever change, which tells us that there has to be something more to the Universe than empty space, since we know the Universe exists, and does so in a form where everything within it is in a constant state of change, with the possible exception of those features which have a role at the subatomic level, which scientists refer to as: *the constants of nature.* So, what could this additional necessity be?

From all that we currently know about the Universe and the nature of things, there seems to be only one candidate that would fill the above requirement and that is *energy,* but exactly what *kind* of energy is open to debate.

It may be the mysterious *Dark Energy,* which in modern cosmological thinking is thought to play a leading role in how the Universe operates, or it may be the so-called *Vacuum Energy* which supposedly exists in space and permeates throughout the Universe. On the other hand, it may be the substance physicists call *Quantum Foam,* which again is thought to permeate throughout the Universe but on a scale far, far smaller than the size of an atom where it is said to exist in a state of constant turmoil - or - it may be something we are currently totally ignorant of.

Either way, the energy in question must be something that underlies every other thing in the universe, that exists eternally throughout the universe, and, most importantly of all, it must be something that does not of itself, have to be created, whilst possessing the ability to influence the creation of everything else.

Conclusion: Ultimately, space is not the empty void

we tend to think it is, but rather we should think of space AS energy with the ability to transform into a physical form, which thereby provides us with what we call *matter.*

That may be a difficult concept to grasp, but as a comparison, think about it like this :

When you seat yourself in a room, you will be concious of the walls and all the objects around you, but you will not be conscious of the fact that you are, in a sense, in contact with everything else in the room by virtue of all the particles in the air and the *forces of nature,* (see appendix A,) which operate in and between the particles.

Likewise, when you leave the room and venture outside into the world beyond, those same connections will remain with you wherever you go. There is no escape.

An easy way to visualize such connections is to imagine you have suddenly shrunk down to a very much smaller scale. The world would then appear as a very unfamiliar environment, with the walls in your room being anything but smooth.

Another way to consider the above, is to think of your physical form as solidified energy, moving through other more fluidic fields of energy, because as Albert Einstein pointed out, mass and energy are simply different forms of the same thing.

THE ILLUSION OF REALITY

If you could suddenly see the world on a smaller scale, everything would become less familiar.

You can still have gaps (space) between objects, but those gaps have various *fields of force* operating within them, for example, gravity and magnetism, and the gaps also represent various dimensions, that is, measures of separation, so everything is far from being a complete and absolute nothingness.
So why is it, you may be wondering, that so many in the scientific community seem reluctant to endorse the notion of a Universe which is both infinite in age and size, and yet so readily happy to accept the illogical assumption that everything in existence originated from Nothingness.
They put forward two arguments. One is related to something they call *Entropy,* and the other is called *The Principle of Causality,* or more simply put, *First Cause.* Let's look at each in turn.

Entropy basically relates to the dispersal of energy into its surroundings whenever permitted to do so, (example : hot water will always flow to cold water,) and since energy is what drives all the activity in the Universe, via the transfer of heat, the thinking is that once that heat is spread out evenly through all that exists and attains what scientists call a *state of equilibrium,* then all the natural processes that are taking place across the cosmos will gradually grind to a halt, resulting in a Universe which is - to all intent and purposes - dead.
Just how long all this would take to come about is

‘undetermined,‘ since the total mass of the Universe is not known, and if the Universe *is* truly infinite in scale, then a state of equilibrium would never come to pass and the Universe would be eternal. Apart from that, even a so-called *dead* Universe is still a Universe and there are, in fact, certain theories which hint at ways the heat death may be avoided.
For a full overview of that : see the Wikipedia article : Heat death of the universe.

A similar counter argument could be applied to the view that everything has a first cause, because if the Universe is infinite in *time,* i.e eternal, you could never reach a first cause however far back in time you went.

It seems reasonable to state, therefore, that the acceptance of the notion that there is no such thing as *absolute nothingness,* nullifies both the Entropy and First Cause scenarios that are touted as objections to a Universe that is infinite in both size and age.

ALTERNATIVE THEORIES

As you might expect, apart from this essay, there are a number of other theories which attempt to explain the existence of the Universe.

One of the most well known is a conjecture called: The Steady State Model, created in the late 1940s by the astrophysicists Fred Hoyle and Thomas Gold, along with the cosmologist/mathematician Herman Bondi, principally to explain the cosmic expansion, without the need for a so-called Big Bang.

In their scenario the Universe is expanding out to infinity, whilst the amount of matter in space, per unit volume, remains constant by the spontaneous and continuous creation of new Hydrogen atoms from within the very vacuum of space.

This effectively meant that the Universe had been, and would continue to be, in existence forever. In the decades to come, however, new astronomical knowledge relating to the distribution of celestial objects known as Quasars[2] and Radio Galaxies[3] revealed a disparity when applied to the claims of the Steady State Model, for the celestial objects could only be observed at great distances, as opposed to *all* distances as posited by the Steady State Model.

As a result, by the early 1960's, the Steady State Model fell out of favour with most cosmologists, leaving the Big Bang Model to reign supreme. It did *not, however,* put an end to the Steady State Model concept, for Fred Hoyle, together with theoretical astrophysicist Geoffrey Burbidge and astrophysicist Javant V Narilkar, produced a new version of the model, in 1993, renaming it: The Quasi Steady State Hypothesis, which still ultimately saw the Universe as both infinite and eternal.

In an article entitled: 'A Quasi Steady State Cosmological Model, with creation of Matter,' as published in the Astrophysical Journal, Volume 410, page 437, in lines 4 & 5 of the 'abstract,' they suggested that the new material which appeared in the Universe, came into existence via "little big bangs," which thus maintained the density of space (per unit volume) throughout all time.

Although the scenario was not accepted by all, its legacy has endured to the degree that even at the present time, there are those amongst the scientific community who still feel the hypothesis warrants some attention.

Another alternative to the 'Standard Model of Cosmology,' (which is what the leading Big Bang model has become,) is the Oscillating Universe model, which, as the name suggests, posits the universe as being subject to a continuous recycling process, alternating between a period of expansion, followed by a period of contraction before a new period of expansion and so on.

That basic concept is thousands of years old, indeed it can by traced back through history to The Hindu Rigveda, an ancient collection of Sanskrit hymns, believed to have been written in India between the 15th and 12th centuries BC, which refers to the Brahmanda[4] or Cosmic Egg, said to be the repository of everything in existence, which expands and contracts from a single point called *a Bindu,* (a Sanskrit term meaning point or dot,) with the whole process repeating endlessly.

Its more modern equivalent,[5] in terms of the Oscillating concept, is said to have been formulated in the early 1920's by the Russian mathematician Alexander Friedmann, and later, in the same decade, by the Hungarian physicist, Cornelius Lanczos, which, at the time aroused much interest.

In the 1930's, however, the work of the American Richard Tolmann, a mathematical physicist and physical chemist, suggested that the Oscillating Universe Model would not be viable because of the 2nd law of Thermodynamics, which basically states, that the efficiency of a system, is predisposed to deteriorate with the passage of time. Scientific opinion was thus again prompted to favour the Big Bang Model, but like the Steady State Cosmological Model that came before it, the modern version of the Oscillating Universe Model still retains some support, since the Richard Tolmann argument against the viability of the Oscillating Universe Model is germane only if you accept that the Universe is finite.

Needless to say, there are a great many other alternative suggestions for cosmological models of the Universe, but there is one thing they all have in common, just as this present essay does, and that is, they are all forced to make certain assumptions, which are difficult or impossible to validate at the present time.

That said, the view of the present author is that the Universe may well alter radically over time, compared to that which we see today, but it will never reach a point where it ceases to exist totally, for that would result in an absolute nothingness, which goes against the logical deduction that there is *no such thing as absolute nothingness,* for if such were otherwise, then the universe would never have existed.

FINAL THOUGHTS

Although the primary purpose of this essay is to promote the notion that the whole Universe could NOT have arisen from absolute nothingness, because there is *no such thing as absolute nothingness* except as a concept, there are some in the scientific community who will try to make a case that favours the opposite view by announcing, in earnest, that there are (apparently) four scientific definitions for *nothing,* all of which are said to be valid depending on what terms you use for such to be fully understood. (*see ref 6, for a link to more info.*)
Two of them are said to permit the birth of the Universe from nothing, with the principle one relying on the notion that *empty physical space* has the necessary energy to bring about the said *creation of the Universe,* which can then evolve thereafter, in the manner that can be reasonably well described with all the knowledge that we currently possess about particle physics.

Unfortunately, the problem for most of us with the above scenario, is that if you have an energy source in 'empty' space, supposedly in a time when there was no Universe, that counts as something that pre-exists the birth of the Universe, (however you want to define it,) and thus it would mean there was a Universe (of a kind) before the birth of the Universe - which is clearly a contradiction.

Choosing to define what absolute nothingness is, in a different way to the norm, is not really resolving the question: is the Universe finite, or infinite and eternal? because the solution attained is in conflict with logic.

Overall, therefore, and on balance, for the present author at least, the notion that the Universe is both infinite and eternal answers more questions than a finite Universe. Whether that is an indication that it may be nearer the truth of the matter, only time will tell.

APPENDIX A

THE FORCES OF NATURE

GRAVITY
The force of attraction between bodies with mass.

ELECTROMAGNETISM
The force of attraction or repulsion between charged particles. It also keeps electrons in orbit around atoms.

THE STRONG NUCLEAR FORCE
The short range force which binds the protons and neutrons together to form the nucleus of atoms.

THE WEAK NUCLEAR FORCE
The very short range force that operates within protons and neutrons. It maintains the correct amount of energy in the nuclei of atoms, by permitting protons to convert to neutrons, and vice versa, whilst releasing, electrons, positrons and neutrinos from the atomic nuclei.

NOTES & REFERENCES

1) Although Edwin Hubble is most often linked to the discovery that the Universe is apparently expanding, Vesto Slipher is credited with collecting the first observational evidence.
Source : Coming of age in the Milky Way : Timothy Ferris : 1998 : p206

2) Quasars. Extremely bright and remote celestial objects which radiate with a light equal to that as from the combined output from hundreds of galaxies, if not more.
Source : https://www.thefreedictionary.com/quasar

3) Radio Galaxies. A celestial object, typically an elliptical galaxy which emits substantial amounts of radio energy. Such electromagnetic radiation at radio wavelengths, was first detected from a galaxy way back in 1931, by the American physicist and radio engineer : Karl Guthe Jansky.
Source : An Introduction to Astronomy : 6th ed, p329 :Robert H Baker. 1961.

4) Drawn from the internet based article entitled: 'Cosmological Theories Through History.'
Source : https:// www.physicsofthe universe/dates.html
retrieved : 28 January 2021.

5) Drawn from the internet based article entitled : 'Oscillating Universe Theory.'
Source : Briceno V Gabriela. (2019). Oscillating Universe Theory. Retrieved : 28 January, 2021, de Faqs. Zone: https://www.euston96.com/en/oscillating-universe-theory/

6) How Did The Entire Universe Come From Nothing? Ethan Siegel.
Source : https/www.forbes.com/sites/startswithabang : 27 November, 2020

Made in the USA
Monee, IL
04 November 2026

67374706R00017